GEOLOGICAL HOWLERS
Boners & Bloomers

Edited by W. D. IAN ROLFE

Hunterian Museum, Glasgow

Drawings by David L. Dineley and others

London
GEORGE ALLEN & UNWIN
Boston   Sydney

© The Geological Society of Glasgow, and W. D. Ian Rolfe, 1980
This book is copyright under the Berne Convention. No
reproduction without permission. All rights reserved.

**George Allen & Unwin (Publishers) Ltd,
40 Museum Street, London WC1A 1LU, UK**

George Allen & Unwin (Publishers) Ltd,
Park Lane, Hemel Hempstead, Herts HP2 4TE, UK

Allen & Unwin Inc.,
9 Winchester Terrace, Winchester, Mass. 01890, USA

George Allen & Unwin Australia Pty Ltd,
8 Napier Street, North Sydney, NSW 2060, Australia

First published by the Geological Society of Glasgow in 1980
This edition published in 1983

ISBN 0 04 550032 0

Printed in the United States of America

CONTENTS

	Foreword ...	iv
1	As General Hutton said ...	1
2	Kragon tails. ...	7
3	The probable cause of earthquakes. ...	12
4	Of boarholes and aquifas. ...	14
5	Pyrites mistaken for fools gold. ...	18
6	Igneous lamentation. ...	21
7	Volcanoes are normally quite gentle. ...	24
8	Schists of better grade....	27
9	Sedimentation is a rather lengthy affair. ...	29
10	From underneath the microscope much can be learned.	33
11	The first muddy division is the Oxford Clog. ...	35
12	Dip horizontal, strike unknown. ...	39
13	Phossils....	42
	Contributors. ...	53

FOREWORD

To the jaded examiner toiling through a pile of geological examination scripts, a good howler seems an oasis. His mind is refreshed, a flight of fancy may be indulged in, before he resumes his task. This volume is the fruit of the oases, collected over many years by geology examiners around the world. I am indebted to the contributors who noted these howlers, and took the trouble to send them to me, thus rescuing them from oblivion. Such contributors are identified by their initials which follow each howler, and their names are listed at the end of the volume. Almost twice as many howlers were submitted as have been used, and I apologise to those who find their gems omitted. Senses of humour do vary, of course, and therefore I have tried to include at least some howlers which obviously appeal to others, although they leave me cold.

Howlers are also known as "boners" in the United States, and Webster's dictionary actually cites a geological example by way of illustration, — "such as dinosaurs surviving until medieval times". "Bloomers", a few of which are incorporated here, are more embarassing, usually Rabelaisian, blunders (see Gathorne-Hardy 1966, Anon. 1967). Some howlers pass into deliberate facetiae, e.g. *Aquatarts (fossil mermaids used to date the Dalradian)*, but these have been included nevertheless. It is but a short step to such published efforts as Haile's spoof (1977) on *Ozymandias,* Stümpke's (1961) *Rhinogradentia,* and T. Hamaiashi's 1975 (for 1932) *Eocene agnostid with 9 thoracic segments* (shown to me by Dr. J.K. Ingham). There is a very delicate path to be trodden here, for such howlers can develop a life of their own. Thus the flightless Ephemeroptera (mayflies) reported in 1978 from the Auckland Islands also only exist in the imagination. They apparently arose "by saltatorial macroevolution caused by the infamous mutagen, the *lapsus calami* (Hubbard 1979).

Some howlers may inadvertently open windows onto new lines of thought. As Freud (1960, pp. 59-60, 93) pointed out in his least known work, the apparent stupidity of a joke may conceal a deeper question: "sense in nonsense". Thus Lichtenberg's joke "He wondered how it is that cats have two holes cut in their skin precisely at the place where their eyes are", neatly raises a problem that only became explicable later, by Darwin. Other howlers may even serve an educational purpose, by drawing attention to weak points in some teaching, e.g. of the magnetically striped ocean floor. Not, I hasten to add, that there is any such weighty purpose behind the present compilation. Far more utilitarian to regard these howlers as a source of pure pleasure, as "a relief from the compulsion of criticism" (Freud 1960, p. 127).

My greatest debt of gratitude obviously goes to the perpetrators of these howlers, who must remain anonymous. As Anderson (1970, 1977) has pointed out, the student's words should be reported accurately, with nothing altered or added in any way — and misspellings have therefore been retained where they help the howler along. I have followed H. Cecil Hunt's advice in his 1928 volume *Howlers:* "In the specific sections, howlers touching on the same subject have not necessarily been brought together. It was found that by scattering them haphazard, the full benefit of their unexpectedness was secured". I am grateful to Ernest Benn Limited for permission to quote the geological howlers from Hunt's volume. An earlier howler is cited here (p. 47), collected by no less a person than Marie Stopes in her 1910 British Journal of Botanical Humour *The Sportophyte,* edited by "the palaeophytologist of Manchester assisted by a flare of northern lights". I am indebted to the current palaeophytologists of Manchester, Joan Watson and John Franks, for providing access to these rare volumes of esoteric humour.

Other geological howlers appear in the series of volumes edited by Denys Parsons (1952, 1953, 1954, 1958). I am grateful to the author and to Macdonald and Jane's Publishers for permission to quote them here, and to them and Peter Kneebone for permission to use the cartoon which appears on page 3. Professor D.L. Dineley kindly contributed all other cartoons, except for those appearing on pages 22 and 27, which are by Peter Carr, page 32 by Graham Taylor (after sketches by the contributor) and pages 7, 8, 31 and 41, after the original authors. Anne Rodger kindly designed the title-page and cover, and editorial assistance was provided by Eleanor Black, George Donald and Graham Taylor. *Geotimes* has long published geological humour, and I thank "Sandstone Sam" and the editor, Wendell Cochrane, for permission to cite his collection of howlers. B.W. Anderson, and the Gemmological Association of Great Britain, gave permission to quote excerpts from his articles (called to my attention by Mrs. J.W. Thompson), although the quotes are not as fully in their setting as he would have wished.

All profits from this volume will help the more conventional publication programme of The Geological Society of Glasgow. I would be grateful to receive further howlers for a possible second edition of this volume, including "behavioural" howlers, such as that which follows. This may also serve as a closing parable on the difficulties of selecting howlers for publication! I was once called to the house of a local enthusiast who pressed me to see his collection. The contents of suitcase after suitcase, box upon box, were carefully unwrapped, and their locality labels read out to me. This huge collection consisted of nothing more than pebbles and cobbles of various, not particularly interesting, shapes. Independently, the collector had evolved a classification of them into "soft fruits" and "hard fruits". From the wide distribution of cobbles which he identified as fossil coconuts, he deduced that Scotland went through a desert period in the not too distant past. My normally low reserves of tact were tested to the utmost in trying to patiently explain that none of these was a true fossil. After several hours of viewing his collection, we parted, each convinced of the folly of the other's views.

REFERENCES

ANDERSON, B.W. 1970. Examiner's rewards. *J. Gemmol.* **12**, 61-4.
    1977. A few more "rewards". *J. Gemmol.* **15**, 345-6.
ANON. 1967. *A new garden of bloomers.* Bodley Head, London.
FREUD, S. 1960. *Jokes and their relation to the unconscious.* In Strachey,
    J. (ed.) *The standard edition of the complete psychological works of Sigmund Freud.* Hogarth Press, London. [Penguin Books, 1976.]
GATHORNE-HARDY, E. 1966. *An adult's garden of bloomers.* Bodley Head, London.
HAILE, N.S. 1977. Preparing scientific papers. [Columnar rock structures from an antique land.] *Nature, Lond.* **268**, 100.
HAMAIASHI, T. 1975. An Eocene agnostid with 9 thoracic segments.
    [Privately printed as *"Kyoto Biol. Rep.* **3**, (9), 1-4'*]
HUBBARD, M.D. 1979. On the origin of flightlessness in Ephemeroptera. *Syst. Zool.* **28**, 227.
PARSONS, D. 1952. *It must be true.* Macdonald, London.
    1953. *Can it be true.* Macdonald, London.
    1954. *All too true.* Macdonald, London.
    1958. *Many a true word.* Macdonald, London.
STOPES, M. 1910. *The Sportophyte* **1** (1).
STÜMPKE, H. [=STEINER, G.] 1961. *Bau und Leben der Rhinogradentia.* Fischer, Stuttgart. [=1962. *Anatomie et Biologie des Rhinogrades.* Masson, Paris.]

1.
AS GENERAL HUTTON SAID ...

The average person does not have to dig a deep hole to remind himself of the past.   C.D.G.

General Hutton said, "The present is the key to the past."   C.D.G.

At one time Wales was a steaming jungle.   J.G.Mc.D.

The enormity of geologic time.   D.L.D.

The theory of continental drift was first propounded by Weber.   D.L.D.

The rock just appears to disappear.   J.H.McD.W.

If a suspected criminal claimed he was in one place when he was supposed to be in another, a heavy mineral study of the clay and sand on his boots would soon prove who was right. Unfortunately, most criminals would clean their shoes.   J.H.McD.W.

We have come a long way since Wagner proposed a flight from the poles.   M.Br.

The North Sea is salt because of the Yarmouth bloaters.   H.C.H.

The dating of rocks depends very much on the superstition principle.   B.E.L.

Sexontidised maps enable you to locate your position more accurately.   B.E.L.

When India collided with the Asiatic block, the Himalayas were formed; when Africa merged into Europe, the North and South Downs were formed.   J.A.

Kaolin is metamorphosed feldspar, very hard and resistant, frozen stiff during the ice age and never recovered since. It is found in moraines and burns the fingers because it is so hot. That is why it is used in poultices. *T.N.G.*

Geological collomb. *A.R.Mc.G.*

If you see a big boulder that looks as if it came out of the sky, the chances are that it is an erratic. *B.J.B.*

[*Evidence of unconformity*]... In a map, the contour lines cross one another. *G.B.*

Just bear rock. *J.H.McD.W.*

A change in basic rock time. *J.H.McD.W.*

Sumposium. *J.H.McD.W.*

Atoll lagoons can be up to 20,000 feet deep. *C.J.B.*

Glaziers are common, they move about one foot per day in Switzerland. *H.C.H.*

Frank confessions of ignorance are often rather appealing, as in the answer "The colouring matter of emerald has been forgotten, but the writer does not think it is manganese", while one sympathizes with an unfortunate overseas student who had obviously been cramming for dear life, and broke down in mid-stream, with the remark "I have learned too much too quickly, and can no more think of nothing at all". One student notion, however, that the singular of specimen must clearly be speciman deserves mention. *B.W.A.*

The specimen shows a strange spiralling shape. The mystic spiral has amazing and deeper meanings thought .... to be the central point of life. It is interesting that early forms of life .... should show distinct signs of spirals. All very interesting but nothing to do with this question. *W.W.*

Glaciers spread a murrain over the land. *H.C.H.*

According to the experts, computers and Jehovah's witnesses we are about to have another ice age. *C.P.H.*

The limestones of which the Houses of Parliament are made are changing to Dolomite and this is causing Parliament to dissolve. *C.P.H.*

I would like to think that the English and writing in no way detracted from the understandabily of the report, and that any terminology used was adequate relevant. Thank you for reading this report. *C.R.R.*

The core has large amounts of iron and knickel in it, so earning it the name of the knife.
    I.M.S.

Doldrums are a series of high rocks near the Equator.
    H.C.H.

Beginning in the winter of 1938 Dr. Ewing and his associates, working on the deep-sea research vessel Atlantis, began to experiment with underwear photography.
    D.P.

Hutton said that the present was the result of the past, which he called Uniformitarianism. Verner, on the other hand, preferred to believe in phenomena like Noah's Ark and said that granite was a sedimentary rock.
    M.B.

Isostasy is a principle of equilibrium which was supposed to have been discovered in Archimedes' bath.
    W.G.J.

When two suites of rock lie unconformably on top of each other there is absolutely no relationship between the rocks. *W.G.J.*

Geochemical studies in and around the south-east corner of the Tavern window. [*D. Phil. thesis*] *Ga.B.*

The Earth isn't all it's cracked up to be. *J.H.*

Pinnate water. *J.A.W.*

All the chemical elements are dissolved in sea-water. The explanation is that rivers have been carrying dissolved miners into the sea for millions of years. *D.P.*

The continents are a triple junction between aerial, surface and marine zones. *C.P.H.*

Hot spots occur near to seduction zones. *C.P.H.*

The land was uplifted probably due to William Smith's ideas on laws. *I.M.S.*

Mountains are formed in erogenous zones. *C.J.B.*

Two theories of isostasy have been expostulated. *W.E.T.*

Colition of Continents. *J.H.McD.W.*

On an advertisement for a plate tectonics short course being offered to industry, the heading "continental rapture". Assuming that it was related to seduction zones and erogenic beltsand might even have something to do with "the origin of divine basalt". *I.P.*

Bloke diagram of island arc. *J.H.McD.W.*

The term Calodonian Orogeny is brandished about by many people.
J.H.McD.W.

A geosyncline has a great effect on the typography of the land. *A.R.Mc.G.*

I also use common sense in assuming that the land wants to spread a bit — this also gives the ocean a chance to spread about also. *N.J.H.*

Plates on the Earth's surface are driven by infernal processes. *C.P.H.*

The crust is destroyed by a series of sink holes. *W.G.J.*

The means by which continents move apart is best illustrated by Vine and Matthews when writing on the mid-Atlantic ridge. *C.P.H.*

When 2 plates are pulled apart larvas come to the surface. *C.J.B.*

Orogeny recapitulates phylogeny. *R.L.B.*

Subjunction Zones [*7 times in one essay!*] *J.H.McD.W.*

Depressions may have their bottoms quite a distance below sea level. *J.H.McD.W.*

... the rock will become less thicker. *J.H.McD.W.*

In the lab. no substitute for time has been found. *J.H.McD.W.*

A breccia may be formed by desiccation (dry rot), or by frost action (cold rot), or by solution (wet rot), or by the action of the sun (hot rot). It may happen by faulting in any kind of rock (all rot). *T.N.G.*

If it is found that coal is not found .... *J.H.McD.W.*

There are three kinds of rocks, ingenious, sedentary and metaphoric. *M.B.*

Scotland must have been much thinner. *J.H.McD.W.*

In 1620 Bacon noticed the fitness of the continents. *A.R.McG.*

Man — appendix — was a tail at first now is in mans stomach as an appendix. *N.J.H.*

The large comet was seen by a resident in the heavens in the direction of the Forest. *D.P.*

Sometimes they are confined by bosses. This is known as "confined virgination". *J.H.McD.W.*

The petrifying of stones in springs can also be seen in the household kettle. *J.H.McD.W.*

Nine-eighths of an iceberg is beneath the sea. *C.P.H.*

If the geology of a site is not studied it may not be noticed that it is in the middle of a geosyncline and thus the plans may not be altered accordingly. *C.P.H.*

The Alpine orogeny came about because of Italy crashing home into Southern Europe and displacing the geosynclinal sea which is known as Tethys. *C.J.B.*

Many orogenies began early in some other part of the world and only affected Britain an era or two later. *C.J.B.*

An orogenic movement brought to a close one section of the stratigraphical column. *C.J.B.*

The Rockies grow because of the Theory of Isostasy. *C.J.B.*

Plate-Tectonics has been called the Uniforcation Theory of geology.
C.J.B.

Herculyean movements. *G.B.*

Laplace said there was a mass of gas just standing around doing nothing. *T.C.F.S.*

Crush in a mortar and pistle. *J.H.McD.W.*

The Earth makes a resolution every twenty four hours. *H.C.H.*

Lyell and Darwin proved the Biblical story of the flood to be untrue, and the religious people who followed their teachings were called the Unconformitarians. *T.N.G.*

The first law of geology is the law of supposition. *D.K.B.*

It has been postulated that Iceland is still experiencing the Tertiary Period. *I.S.J.*

Hade is a deep hole in the Earth. *G.W.T.*

The continents on Earth have relatively free movement due to the large percentage of water, whereas if the whole surface was solid then this would involve a tremendous force and a great deal of overlap of these rocks. *D.L.D.*

Party to leave bus station, bus No. 18. Alight at Sea Corner, Highcliffe, and proceed along shore. Tea at Barton-on-Sea. Small chisels advised.
D.P.

We are not amused (Duke of Gondwanaland). *P.A.*

A glacier is something in Australia that explodes. *H.C.H.*

The Midlands (which are an extension of Wales) .... *C.P.H.*

There are two possibilities .... Since the latter is more modern and since the former involves more, I will talk about the former. *C.P.H.*

The Earth would have taken a long time to cool if it had not been for ice-ages during the Pre-cambrian. *C.P.H.*

2.
KRAGON TAILS

A raised beach may be formed by submergence of the sea.   *G.B.*

The Avon is the only British river to show great superimposition and this is due to rapid downcutting in Devonian times.   *G.B.*

This forces the crack to widen until it falls off.   *G.B.*

Superimposed drainage is seen to occur when the direction of flow of a river bears no relation whatsoever to the topography of the surrounding countryside.   *G.B.*

Whenever a land rises out of the sea or vice versa ….   *G.B.*

…. carrion power of a river   *G.B.*

meanders   *G.B.*

The water is got further up the river by corrosion and transported down the stream either in suspension on being rolled along the river bed.   *G.B.*

Boulder clay speaks for itself.   *G.B.*

The Nile has an accurate delta.   *M.B.*

The fresh snow melts on the ice and works its way through it, pushing the ice further away all the time and forming a hole between the rock and the ice in which the water then accumulates. Later the ground moraine at the bottom relaxes and bends upwards and crevices are then found in the ice. [*Answer to a question on the mechanics of glacier erosion*].   *M.B.*

A glacier has passed through the area, cutting off the landscape.
    J.H.McD.W.

Antecedent drainage is that which flows across the tops of folds from the anticline of one fold to the anticline of the other.   *C.J.B.*

A mortlake is formed on a glacier — there it is when the ice melts, a 'dead lake' left high and dry up in the air with only the ice beneath to hold it up without a chance of its getting away.   *T.N.G.*

The Thames drainage basin:—

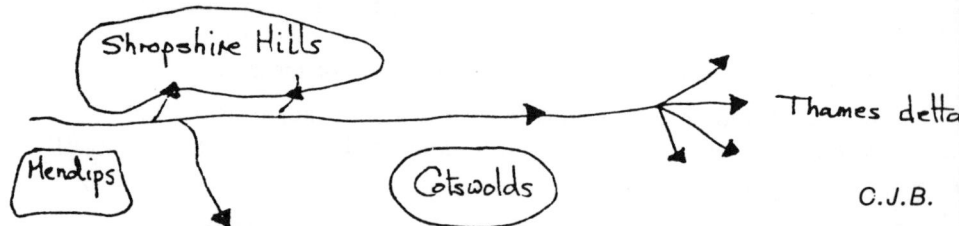

C.J.B.

In the deserts of Mexico you can find water in shallow wells in the little delta fans at the foot of the mountain ranges, and when the first Spanish explorers discovered them when they were dying of thirst they were very happy and so called them bahahadas. *T.N.G.*

Foliage is leaves, so exfoliation is the removal of leaves by ice action. This is known as onion weathering because onion trees are very easily attacked by ice. *T.N.G.*

Exfoliation is caused by cloudbursts. The French call them oranges because they peel off. *T.N. G..*

An esker is a ridge that runs across the country, sometimes over the horizon. It is formed by ice, and when the ice melts the esker is left behind covered by a string of beads. *T.N.G.*

An esker is formed under conditions. These conditions are very exceptional and no-one has ever seen them. Even lecturers don't know how eskers were formed, but there are many guesses. I have never seen an esker and I think it is unscientific to guess. *T.N.G.*

A levee is a result of ice action. This is obvious when you go to Helensburgh, where the shops are built on a glacial raised beach and they charge a high levee to summer visitors. In fact, the whole of the Clyde coast is very topographic and proves the point conclusively. *T.N.G.*

Drumlins are shaped by glaciers moving backwards. *C.J.B.*

Drumlins form a relief which is known as a nest-egg topography. *C.J.B.*

Terminal moraines are formed by a retreating glazier. *C.J.B.*

Esker — a long, thin piece of land emerging from a glacier. *C.R.M.*

An evidence of former glaciation is scared bedrock. *R.L.B.*

The shape of the Lake District is often called an inverted spoon.
J.H.McD.W.

The glacier flew from the east to the west.   C.J.B.

Glaciation occurred in the area leaving a Kragon-tail behind the resistant plug of igneous material.   C.J.B.

When the river reaches old age, the gradient of the land is virtually nil and rivers have to find their way to the sea by some other method apart from flowing in a straight line. So it meanders, i.e. flows from side to side to gain as much speed as possible to reach the sea.   J.A.

The sea is eating into the limestone. A similar feature can be found at Ludlow Cove.   G.F.

Eventually spring broke through causing a radial drainage pattern. C.D.G.

.... the hillside was covered with erotic blacks.   J.S./H.E.W.

Outletless lakes.   J.H.McD.W.

Salts formed by evaporation of dunes.   J.H.McD.W.

Melt water forms a gorge. Such Georges are frequent occurrences in N. Wales.   J.H.McD.W.

Lacustransine.   J.H.McD.W.

Mountains are altitudinous.   J.H.McD.W.

The coombes of the South Downs have recently been invaded by extensive deposits of periglacial slugs.   C.R.M.

Great Britain is loosing about ten feet of coast each year with the effect that Blackpool is getting nearer to Skegness.   I.M.S.

A water table is a record of direction of flow of storm water from the hills into the rivers and streams from which any change such as river capture may be noted.   G.B.

The 920 feet terrace is well given by $y = 1121 - 175 \log(160x) + 1.1x$ where x is measured in sixths of a mile from Totnes railway bridge. N.B.

From Llandrindod you proceed along the lovely valley of the Ithon, growing more beautiful as you proceed.   D.P.

The extent to which the landscape is eroded by glaciers also gives evidence as to where a glacier extended as far as.   J.H.McD.W.

High altitude has a considerable effect on relief.   J.H.McD.W./G.A.W.

Tributaries have made minor breeches in the valley sides. J.H.McD.W.

The river is embedded in the quartzite.   J.H.McD.W.

Erratics ... have been left as isolated rocks which have been allowed to wander. *A.C.McL.*

Mango swamps. *M.F.*

A till is a fossiliferous tillite. *C.P.H.*

Glaciers scrape the valley sides leaving marks known as slick on slides. *C.P.H.*

Drumlins are formed by the formation of dirty ice which moves slower than clean ice and is therefore shaped like an upturned boat. *C.P.H.*

A raised beach is basically just that. Due to earth movements the land has risen, or the sea has dropped, or both. Thus suspended above the sea is the old or raised beach. *I.M.S.*

An ox-bow lake is formed when a river meanders, suddenly reverses and moves in the opposite direction. *I.M.S.*

No other glacial or subsurface features were seen because of the dry stone walls of the area. *C.R.R.*

Gently rooling hills. *J.H.McD.W.*

The Parallel Roads of Rob Roy were made when the level of the lake fell three times as the Highlanders cleared the log jam that held up the water, and so enabled him to escape to his hiding places near the head of the River Spey when he was chased by the Hanoverian soldiers. *T.N.G.*

The parallel reads of Glen Roy are an unparalleled example of this phenomenon. *J.A.W.*

[*Whole school's explanation of river meanders:*] Eight locomotives were pulling flat trucks with extremely long steel lines on them, the locomotives crashed into a bulldozer and the shock of this caused the steel lines to form themselves into meanders to distribute the shock throughout. *J.G.M.*

Fifth Army seizes junction of Parallel Roads. *D.P.*

The origin of water is intimately associated with what is known as the Hydroloric cycle. *C.J.B.*

Superimposed drainage is a drainage pattern of inconsequent nature. *J.H.McD.W.*

As we can all see, as long as the Earth moves a rock will always be formed. *N.J.H.*

Glacial erotics.   C.J.B.

One of the most controversial topics in geology must be the origin of dry valleys.

C.J.B.

[*From a Japanese author to an editor:*] Please thank the referees for their suggestions. I would like to execute them.

R.C.E.

3.
THE PROBABLE CAUSE OF EARTHQUAKES

Vine and Matthews saw a magnetic anomaly at every ridge throughout the world. *C.P.H.*

The continents are being searched for magnetic wanderers. *M.Br.*

Seismographs are machines which pick up the vibrations of an earthquake before it happens. *C.J.B.*

.... from the penis suspended a heavy weight to prevent bouncing. *F.H.*

Earthquakes tell us many things about the Earth — from an earthquake you can sometimes measure the degree of heat coming from the centre of the earth — the flow of lava can produce many fossils, plus also preserve other rocks and substances. *N.J.H.*

In ocean floor basalts the small magnetic metals crystallize in either black or white strips according to which way the pole is. *C.J.B.*

Carbon dioxide allows the Sun's Rays to penetrate through, but on reflection at the Moho discontinuity the heat is prevented from escaping into the upper reaches of the Earth's atmosphere. *C.J.B.*

The rise in temperature with depth can be shown by dropping a thermometer down an old shaft. For every mile the instrument drops the temperature increases by 17°C. *C.J.B.*

Seismic waves increase their velocity as they go downwards but they get very shaky at the 11+ stage. *T.C.F.S.*

Before an earthquake occurs shock waves are given off. *C.J.B.*

Only now are we progressing beyond the earthquake detection of the Chinese — the dragon which dropped one of its balls when the ground shook. *G.P.D.*

$$\frac{\sin x}{n} = \frac{\sin x}{n} = six$$

G.P.L.W.

The Earth is 4,500,000,000 million years old. *C.J.B.*

Iron ions in deposits on the sea floor may have come from volcanoes. These ions were thus very hot. If over the temperature 400/500° C (Currie Point) they become magnetic and move about in the water to settle pointing in the direction of the North Pole. *B.E.L./D.B.*

The probable cause of earthquakes may be attributed to bad drainage and neglect of sewage.
H.C.H.

[*West Country paper:*] Convection currents in the underlying rocks provide the energy and mechanical requirements needed to make the gradual drift or motion of woman's pale green two-piece suit on April 17th.
D.P.

4.
OF BOARHOLES AND AQUIFAS

An aquifa.   *J.H.McD.W.*

Never build your house on a perched water table, because your well is likely to dry up in summer. The difficulty is now disappearing since most modern houses have taps.   *T.N.G.*

boarhole   *G.P.L.W.*

Much of the area is jurassic and crustaceous, the target oilmen aim at when they start to drill as it is the most likely to contain oil.   *G.B.*

Twelve billion dollars' worth of securities are stored in a New York City vault, three floors below the Earth's crust.   *GeoT.*

Permeable refers to the rock's ability to pass water.   *A.D.W.*

The Electrical Industry consumes large quantities of mica from tachylite and picrite in Ardrossan.   *C.D.G.*

The Superior Oolite is used even more than the Inferior Oolite. Many of the Oxford Colleagues are built of it.   *G.F.*

The mineral resources of the area are negligent.   *R.L.B.*

When logging in, do not use the word "log" as a verb. This is an incorrect usage.   *D.P.*

It must be determined as to whether the rocks above will stay up when a mine is built underneath them.   *J.H.McD.W.*

When a normal fault is encountered the machinery is moved forward whereas for the other kind the machinery is reversed before a bore is put down to find the bed again.
G.B.

Kaolin is of course china clay. It is called china clay because it is used for making china. It is also used for making poultices, but it is not called poultice clay.
T.N.G.

Worthless rock associated with ore is called mange.
R.L.B.

During the war the Germans were short of many things so they made many things out of coal including cheese, bread, eggs and things of that nature.
B.J.B.

An oil field consists of one or more related poos.
R.L.B.

These magmatic processes are vital to many aspects of modern life.
C.D.G.

Portland stone is used as a building stone, especially in industry, because it becomes black after a while.
C.J.B.

Cromlechs are fibrous substances found in Scotland, out of which a sort of incombustible cloth is manufactured.
H.C.H.

Kaolin is so well known that everybody has seen it in a poultice, and I shall not waste the examiner's time be describing it in detail. I expect the examiner has seen quarries of kaolin too, and probably could say how it was formed better than I can.
T.N.G.

Buildings subjected to this process of sand blasting take on the three cornered shape known as bierkatter.
G.B.

Boreholes could be sunk at regular intervals a short distance apart. From this you could see a gradual change in the direction you are travelling.
G.B.

Rocks have ability — not all of them — to pass water at high pressure.
G.B.

Long-term stone quarries can be almost geological in scale.
G.L.

Ground water can be obtained in four main ways: the first is from meteorites.
A.R.McG.

Lead used to be used in making batteries because it was heavy and not very portable.
C.P.H.

Quicquid sub terra est .... [*Trans.:*] There's gold in them thar hills ....
W.W.B.

Galena is used among other things in soap manufacture and aircraft manufacture.
                                                                                        D.E.G.B.

Over the range from about 450 degrees centigrade to upwards of 500 degrees centigrade, the coal passes through a phase of elasticity during which it can be moulded between the fingers like putty.   D.P.

Galena's relative economic importance to the everyday person, is illustrated by the constant theft of lead of many church roofs up and down the country.   D.E.G.B.

The main difference between oil and water is that water, meteoric water anyway, falls on to the land and oil doesn't.   J.A.W.

It has been found by a gentle man that organic remains can be converted to petroleum by the processes of metabolism.   G.P.L.W.

Galena is used in the ammunition industry because of its weight which makes it have very high energy when moving. It is therefore very difficult to stop.   D.E.G.B.

Gold is found in nougat in placer deposits.   A.R.McG.

Feldspar can be crushed to be put in several items if needed, for various reasons.   D.E.G.B.

Capitalist will consider financing Canadian oil fields, or will send English theologist to investigate property.   D.P.

Lumps of baryties can be left in a field for the cows to like. *D.E.G.B.*

Limestone is a useful building stone except in towns where acid fumes turn it into Epsom Salts. *J.H.McD.W.*

Limestones are of great economic importance ... some quarries have supplied some £15 worth. *J.H.McD.W.*

Gypsum is the main component of plaster and is therefore greatly used in the building of hospitals. *C.P.H.*

Salt is essential for the well being and consumption of humans. *C.P.H.*

5.
PYRITES MISTAKEN FOR FOOLS GOLD

It is well known, through the use of formica table coverings that mica can be split horizontally but not along the vertical plane.   *M.J.L.B.*

A genuflect twin.   *G.P.L.W.*

The colour of a mineral depends upon the contents of minerals which would give colours like white, yellow, brown, black, grey ... *M.J.L.B.*

The minerals which are soft could be scratched with knives, and if the mineral does not have any scratch, then use a hammer.   *M.J.L.B.*

Micas are pyroxenes.   *M.J.L.B.*

Quartz and calcite look very similar, but a penknife will fell them apart.   *M.J.L.B.*

In testing for a radium-treated diamond, it should be "raped in black paper with photographic film".   *B.W.A.*

Kaolin is a clay formed by the weathering of boulders. Hence it is called boulder clay. Boulder clay is made by ice, and kaolin is a well-known constituent of ice and is put into ice-pack poultices if you have a headache from overwork.   *T.N.G.*

Copper Arsenical Pyrites is called Miss Pickel for short.   *C.J.B.*

Dolorite is a mineral which has a value of 12 when hit with a heavy object.   *C.J.B.*

With true scientific caution, describing a cube of fluorspar — "This mineral is cubic I think; it is purple I think".   *F.H.S.*

Kaolin is easily identified since it is always found with feldspar. Feldspar is either monoclinic plagioclase or its identical twin triclinic orthoclase. This is why it is always difficult to distinguish plagioclase from orthoclase.   *T.N.G.*

Galena has a graphitic lustre and therefore contains lead.   *N.H.*

chalcopyrite: AuO — Gold oxide.   *N.H.*

A crystal is built around three main axes.   *G.P.L.W.*

B is iron pyrites, FeS_2. The sulphur is conspicuous by its gold colour, and the iron by its black, lustrous appearance.   *N.H.*

The specimen has a hardness greater than 5; breaks teeth!   *R.A.H.*

The composition of iron ore is Fe2, and its hardness on the Moho scale is 5.   *C.J.B.*

Iron pyrites is mistaken for Fools Gold by many people.   *C.J.B.*

Zinc Blende is very important, because, once the zinc is removed, it has a wide range of industrial uses.   *C.J.B.*

Lumbago is a mineral used for making pencils.   *H.C.H.*

This mineral has a colourless colour.   *G.P.L.W.*

Flourite is F_2O.   *N.H.*

Blende ... is the illustrious form of haematite.   *N.H.*

Calycopperite.   *N.H.*

What is Kaolin? This is a difficult problem. It is a mineral, since everything in geology is a mineral. You can find out its composition by difficult chemical analysis that only lecturers can do. Where it is formed is something of a mystery. I think weathering has something to do with it, but not much. I wish the lecturers would tell us. I remain baffled.   *T.N.G.*

I find Moh's scale of hardness most interesting.   *B.J.B.*

A crystallographic axis is a line passing through a crystal which determines its preferred orientation in space.   *G.P.L.W.*

The crystal has a diad axis, a triad axis and a gonad axis.   *N.H.T.*

We examined the soil above the quarry and found it composed of biotite mica, which breaks down as do plagio and ortho clays.   *R.L.B.*

Amethysts are in high command.   *B.W.A.*

Diffraction by Little Bragg Angels.   *R.C.E.*

Optised axes (= axes of symmetry chosen in the best or optimum way)   *G.P.L.W.*

An axis of symmetry is defined as an axis about which a crystal may be rotated so that each face occurs more than once in a complete rotation.   *G.P.L.W.*

The mineral is biaxial because it gives a flashy figure.   *G.P.L.W.*

Crystals have tetrad, triad and gonad axes.   *F.E.T.*

M.O. H.'s scale.   *G.B.*

Dome topped pinaquoid.   *B.W.A.*

Crystallographic systems are defined in terms of their peculiar distinguishing features, such as number of folds [i.e. *n-fold symmetry axes*]
G.P.L.W.

Felsbars — include such ores as orthoplades, microprime, elbite and labradorite.
GeoT.

Quartz is a hexagonal crystal, but its composition (SiO_2) is that of glass — it therefore acts as a cubic crystal.
M.J.L.B.

The ultimate test is to lick the mineral: fluorite is not very poisonous.
M.J.L.B.

A salt pseudomorph is a salt that pretends to be amorphous but is really crystalline.
T.N.G.

[*Description of a quartz crystal:*] Habit, prismatic (Soudé Hexagonal). Horizontally striated due to asbestos.
B.W.A.

The streak of a rock is the colour shown when scraped with a pet.
M.J.L.B.

Flourescence suggests an overdose of some ingredient.
B.W.A.

The main mineral centre is at Iron Knob, where copper and tin are found ....
M.B.

For hydrostatic density determinations in water, air-bubbles clinging to the stone should be removed "by means of a camel's hair-brush".
B.W.A.

Kaolin is a white powder that used to be found in Cornwall but the Romans came and took it all away. There is none in the department, it is so rare. This is very unfair I think, because geology should be taught from specimens, and the rule should be, no specimen no question in the exam.
T.N.G.

6.
IGNEOUS LAMENTATION

Igneous rocks contain most minerals.   *C.D.G.*

If no fossils are found, the rock is either PreCambrian or igneous.
C.D.G.

We in Britain tend to think of igneous activity being confined to distant parts — almost an "It couldn't happen here" attitude.   *C.D.G.*

Rhyolite lava moves at 1 m.p.h.   *A.M.E.*

Magmas cooling below the soil level give intermediate rock types.
C.D.G.

When the magma cools at depth, the ratio of quartz to SiO_2 is equal to 45% and 55% to 55% and 45% thus rending the rock acidic. Intermediate rocks have a 35% to 65% ratio and are termed ultrabasic.   *C.D.G.*

The Wallace monument at Stirling is an example of a sill.   *C.D.G.*

Porphyritic lava is oil formed from vegetable matter and porphyites.
C.D.G.

The mineral orgyte is often found in dhykes.   *C.D.G.*

Plutonic rocks are formed from large magmas.   *A.M.E.*

Plutonic rocks are those which have cooled off in the ground.   *A.M.E.*

The igneous hill masses of the central valley [of Scotland] include the Lammermuirs, the Pentlands and the Curfews.   *M.B.*

A collapsed laccolith is called a bismalith.   *C.P.H.*

Placolite has a neck which is badly affected by wind.   *G.B.*

Rhyolite is an extrusive rock formed in igneous instrusions. It is sometimes coarse and sometimes fine in grain. It is often light coloured though it may be dark. It is basic sometimes, but more usually though not often it is acid. It is a common constituent of sedimentary rocks.
T.N.G.

An example of a granite sill is the Winsill which underlies the Penines: it does not appear at the surface and is only found as a result of trial borings.   *J.H.McD.W.*

A transgressive sill jumping through planes of weakness into the next bed.   *J.H.McD.W.*

The hard rectangular basalt was formed in deep seas and its black fire-faced structure makes it very distinguished. G.F.

The pegmatite, which is a fairly hard rock, has a few fossils in it but not many (late Jurassic), it is found in conjunction with basalt and granite. G.F.

Granite is a sst with a fair amount of feldsper rock, cemented by siclia. G.F.

The CO_2 released acts as tiny ball-bearings which lubricate the crystal boundaries. M.J.L.B.

Granite hopperliths appear everywhere. M.J.L.B.

A locolith. P.T.C.

Batholiths are intrusive rocks which are not visibly floored. J.H.McD.W.

Garbbo slitifys on the surface. GeoT.

Dartmoor is a huge platonic intrusion. C.J.B.

Impermeable and permeable rocks can be brought together by an ingeneous intrusion. C.J.B.

Some granites are found nowhere near igneous intrusions. C.J.B.

The Whin Sill of N. England eminates in the Faroe Islands and travels right across the Pennines. C.J.B.

Dykes and sills can be huge or tiny, the latter is often very much bigger than any of the former. C.J.B.

An intrusion of dolerite came up through the middle of the diagram. C.J.B.

The weather conditions changed once more and a layer of granite was laid down. C.J.B.

A visit to an aunt in the North West not far from Carlisle revealed what I think was a dolerite dyke. ConG.

Igneous rock between sedimentary rock is called a shite. G.B.

The coarse grained igneous rocks are called acid igneous rocks. That is because they contain more than 40% quartz. If basic rocks were also included it would mean that the rocks were not coarse grained because here we have one of the fundamental differences between the two. *J.A.*

Floundering of the pluton roof. *J.H.McD.W.*

The whole area [Arran] was then subjected to an attack of dykes.
J.H.McD.W.

Rock salt is basically a very soft igneous rock. *C.P.H.*

These granites are characterised by large initial $^{87}Sr/^{86}Sr$ ratios, suggesting that they are of predominantly crystal origin. *R.M.McI.*

Eventually the sea sank away again, but as it flushed away it left us with a dolerite bed. *G.F.*

The Whin Sill may be found in S. Scotland and North England and is about 30 feet in length. *C.J.B.*

Igneous Lamentation. *J.C.*

Igneous rocks are found in places allocated to them. *D.A.B.*

Intrusion of hot mama. *A.M.H.*

Fogassity. *M.J.L.B.*

Sycamore-tree lopoliths. *J.A.W.*

Towards noon a halt was called and we lunched on some slabs of Kamet's red granite. *D.P.*

7.
VOLCANOES ARE NORMALLY QUITE GENTLE

An Ardente Nuée was experienced at St. Pierre which wipped out all the inhabitants. *J.H.McD.W.*

Mt. Etna, Scilly. *J.H.McD.W.*

Island of Chertzy. *J.H.McD.W.*

Volcanoes have large wide mouths. *J.H.McD.W.*

Gallons of molten lava and boiling mud poles. *J.H.McD.W.*

Calderela. *J.H.McD.W.*

The lack of support for the volcano would mean that it would collapse into the withdrawn area. *J.H.McD.W.*

Pompeii a city in central Italy was completely wiped out by a volcanic eruption in the 20th century. *J.G.McD.*

Krakatoa the 1,000' mountain became a 1,000' hole, in a few minutes.
C.D.G.

A plug is formed in a volcano. The pressure underneath builds up until the plug is finally shot out like a spear, to land quivering in the ground, hundreds of yards away. e.g. Pûy de Dôme in France. *C.D.G.*

Active volcanos are continually expelling lava and hot gases and are not very dangerous in as much as they are known and given an adequate berth by people. *J.G.McD.*

A volcano vomits from the bowels of the Earth. *J.G.McD.*

Mt. Kilimanjaro, a Pacific island, blew itself to pieces. *C.D.G.*

[*Describing a cylindrical core sample of Millstone Grit:*] This is part of a volcanic plug. *F.H.S.*

When the volcano has built up sufficient steam and gases in its bowels, these gases and steam and molten material are forced to the outside through the blowhol. *G.B.*

Molten lava flows out of the Mid-Atlantic rift from time to time and covers the surrounding sea floor with sea floor spread. *C.J.B.*

An example of a great Caldera is the formation of Lough Neogh, Co. Antrim where the part of the volcano that occupied that space was blown into the channel to form the Isle of Man. *J.A.*

Sinter is blasted out of a volcano, blown to bits, and never seen again. The particles are very fine in grain because of the volcanic explosion, and their absence can only be discovered under the microscope. This was one of the most important discoveries made in the nineteenth century and established the science of petrology, but even now there are very few experts who have ever seen sinter. T.N.G.

Greenland Volcano in Eruption
By arrangement with *The Times* D.P.

Peléan eruptions blow volcanic bombs 30 miles into the air. C.J.B.

The explosion of Krakatoa was heard as far away as Australia, it blew windows out in Bolivia and the force knocked a man off his bicycle, cycling on the east coast of India. C.D.G.

If it's a big crate, it's known as a caldera. J.H.McD.W.

The Stromboli type is quiet and noisy. J.H.McD.W.

On a touring holiday in Italy, being a geologist, I just had to go a little out of my way to see the entrails of Mount Etna. ConG.

Pumice is used as an abrasive, often in bathrooms. A.M.E.

The lava has flown out from fissures. J.H.McD.W.

A volcano can extrude just gasses and therefore it will only be a hole in the ground. *I.M.S.*

The dolerite dyke indicates the presents of volcanic activity.   *J.H.*

Acid lava is not very vicious and therefore does not flow easily. *I.M.S.*

The terrain in question is the product of vulcanization.   *R.L.B.*

Gas holes in lava are called vestibules.   *C.D.G.*

Water and gas is a lot more fluid than a solid.   *J.H.McD.W.*

The gas is cooled and evaporates to water.   *J.H.McD.W.*

Kaolin is a product of granite in the explosive phase, carried by wind as igneous dust to be deposited as loess especially in China. Hence, it is called China clay. It is used in beauty culture and has an exquisite smell.   *T.N.G.*

A volcano is a mountain where the world burst through, and the crater is where it spits out.   *H.C.H.*

A sequence of basalt lavas yielded apes 11.8 million years old.
    M.J.L.B.

Volcanoes throw out saliva.   *H.C.H.*

The most common type of volcano is the compost.   *C.D.G.*

A type of material ejected in volcanism is larva.   *R.L.B.*

Volcanic tuffs fall into 3 main categories.   *J.H.McD.W.*

Basic volcanoes are normally quite gentle.   *J.H.McD.W.*

8.
SCHISTS OF BETTER GRADE

Metamorphic rocks can be found as heavy masses.   *M.J.L.B.*

This olivine then gets into the marble and appears as black spots.
    M.J.L.B.

These intrusions have caused scree metamorphism.   *G.F.*

The granite intrusion has a boulder clay aureole.   *G.F.*

The mica schist is one of very common knowledge.   *C.P.H.*

Boulder clay, an acreacerus rock, has intruded the granite. The area then undertook regional metamorphism and deposited pegmitite, a metamorphic rock.   *G.F.*

Schists of better grade.   *P.T.C.*

When mudstone is changed by thermal metamorphism it becomes clay, sandstone becomes shale and limestone becomes various forms of coal.
    J.G.McD.

The limestone was metamorphosed to produce Goniatites.   *C.J.B.*

If a rock has been metamorphosed the radio-metric date will be inaccurated.   *C.P.H.*

When sandstone is subjected to regional metamorphism the weaker and smaller grains are eroded often exhibiting steep valley sides and narrow valley bottoms.   *J.G.McD.*

During metamorphism, water is squeezed out of rocks: such rocks are said to be incontinent.   *W.S.P.*

Highly unmetamorphosed rocks.   *G.B.*

Conglomerate is a finely blended metamorphic rock.   *G.F.*

Schistosity is the ultimate in metamorphism.   *I.M.S.*

Attempts at porphyroblast *by twenty-two students from the University of Waterloo's Class of 1971-2:*

PAPHYROBLAST	PORPHROBLAST	PORPHYROBAST
PORPYROBLAST	PORPHERABLAST	PORPHROPCLAST
PORPHRYOBAST	PORPHYOBLAST	PHORPHROBLAST
PORPHORBLAST	PHORPHYROBLAST	PHORPHEROBLAST
PHOROCLAST	PHORPHROCLATS	PORPHROCLAST
PHORCLAST	PHORPHOCLAST	PHORPHROCLAST
PHROCLAST	PHROPHROCLAST	PORPHRYOBLAST
PORPHYROCLAST	PARPHROBLAST	PORPHOBLAST
POPHYROBLAST	PORPHOROPLAST	PHORPHOROBLAST
PORPHONOPLAST	PORPHOROBLAST	PORYEROBLAST
PORYROBLAST	PROPHYROBLAST	BLASTOPHRYRES
BLASTOPORPHYROCRYSTS	PROPHYROS	GREAT JUMPING METACRYSTS

E.C.A.

Metamorphic grade refers to the change in the amount of change in a rock as the cause of the change is approached. *C.D.G.*

Thermal metamorphism is the action of the sun in breaking down rocks. Regional metamorphism, on the other hand, depends upon specific region: polar, equatorial etc. *B.J.B.*

The area is very old because the shales contain oysters. The shales have been metamorphosed from limestone. *G.F.*

9.
SEDIMENTATION IS A RATHER LENGTHY AFFAIR

The sediment was deposited in deep water faeces. *C.J.B.*

Mudstone is a shale formed by the deposition of silt. *B.J.B.*

A sea-floor dwelling environment for organisms is termed 'Bentonic'. This is because clay is the last part of any sedimentary assemblage to be deposited and the sea bed is always covered in clay, specially the weathered pyroclastic product called Bentonite — hence Bentonic. *M.B.*

The nearer the beds of rock are to the sea, the narrower their wide is.
C.J.B.

Sedimentalisation. *J.H.McD.W.*

Coarse-grained detrital rocks are in the Pshitic category. *C.J.B.*

Currant bedded sandstones. *J.H.McD.W.*

Tiny ools. *J.H.McD.W.*

Ooses. *J.H.McD.W.*

Pyritous shale — shale which has had pyrites. *P.T.C.*

Un-natural sediments such as sunken battleships ... *J.H.McD.W.*

A pair of laminae representing a year's sediment is termed a vulva.
R.L.B.

Symmetrical ripple marks are produced by alternating currents. *R.L.B.*

Sedimentation is a rather lengthy affair. *W.D.I.R.*

Ribble marks. *J.H.McD.W.*

Conglomerate is associated with sand blasting and spotty dick.   G.F.

Nipple-marked sandstone. [note by chief geologist in margin: "?mermaids' wallow?"]   G.T.

Boulder clay is formed by clay soil being compressed to make boulders.   G.F.

Chert usually keeps company with flint. You can often find them hand-in-hand on the down. They are very tough and resist attack.   T.N.G.

Chalk formed in calm, walm waters.   J.H.McD.W.

Foreset beds are old beds, set down 'fore others.   T.N.G.

The sea digressed and laid down a bed of oolite.   C.J.B.

Coal is formed under anaesthetic conditions where the bacteria cannot function.   C.P.H.

The highest ranking coal is Andesite.   C.P.H.

The various stages of coal-formation, e.g. peat lignite-anthracite are strictly controlled by the British government.   C.P.H.

Turbulence causes chaos.   J.A.W.

A tail-less breccia.   J.H.McD.W.

There is a decrease in lithology downcurrent.   J.H.McD.W.

The grits and sandstones built up and initiated the great Armorican Orogeny.   J.A.

There are four types of sapropelic coal: cannel coal, bogside, anorthite and oilshale.   A.R.McG.

A limestone with little plastic sediments.   A.R.McG.

Limestone was formed on a warm shelf.   *C.J.B.*

There is a very contented structure which may be due to bedding.
J.H.McD.W.

Tool marks.   *M.F.*

Rock salt is very soluble and it exists in large deposits both above and below the ground.   *C.P.H.*

There are several kinds of beds in a delta, foreset, backset, downset, and upper set.   *T.N.G.*

Grains transported in solution by rivers are not well rounded.   *N.H.T.*

Greywhackee.   *J.A.W.*

Rhythmic sedimentation is found when neither marine nor non-marine conditions are dominant.   *N.H.T.*

Dust is mud with the juice squeezed out.   *H.C.H.*

Turbites and Turdites.   *J.H.McD.W.*

Pebble stacking (or embrocation). Gutter castes.   *J.H.McD.W.*

Basically, sole structures indicate the tops of beds.   *J.H.McD.W.*

These sediments were laid down by a large river which debauched north of Southampton.   *A.R.McG.*

Mudstone is mainly composed of sand and gravel.   *C.J.B.*

Red Sandstone: although this rock is not extensively found in large areas it does pop up in certain places.   *W.G.J.*

Coal is decayed vegetarians.   *H.C.H.*

Abnormally large oolites — known locally as cannon-ball oolites — are found near the Yorkshire coast.   *A.R.McG.*

Oolites are the shells of small animals that rolled about over the sea floor.   *N.H.*

Guano is the product of manurous birds.   *H.C.H.*

Submarine slide blocks are generally very long in length but extremely thin in thickness.   *J.H.McD.W.*

Sauce of sediments.   *G.P.L.W.*

The largest concretions occur round Burton Bradstock. *J.H.McD.W.*
Palaeocurrent directors. *J.H.McD.W.*

10.
FROM UNDERNEATH THE MICROSCOPE MUCH CAN BE LEARNED

Rock specimen A is oolitic limestone. Its texture can best be described as oolitic. The rock consists of the small ooliths, thousands of them, cemented together so the rock has the texture of lots of little small ooliths, i.e. oolitic texture.
I.M.S.

From underneath the microscope much can be learnt. *J.H.McD.W.*

[*Hand specimen identification:*]
1. It has a light density, *i.e.*, it is not as heavy as it looks.
2. It is a dense rock, *i.e.*, it is heavier than it seems because of its size.
3. It has an average density, *i.e.*, it does not seem heavier than it actually is.
D.G.S.

[*Conversation overheard in an Edinburgh teaching lab:*]
Student: "It looks like a sandstone to me".
Demonstrator: "Nonsense! Look at it carefully with a hand lens and you'll see it's quite different".
Student (examining specimen): "The only difference I can see is that it's now a lot bigger!"
K.C.

Greensand — as its name suggests — is a green sand.   *J.H.McD.W.*

Rock has smell of a muddy sedimentary rock but nevertheless is igneous.
N.H.

Flint is a siliceous organic deposit composed of calcite. It is of sedimentary origin, the silica being common in igneous rocks. It is formed by concretionary nodules permeating through the rock and collected by sponges.   *T.N.G.*

Greensand is a clay. It is formed in the usual way that sands are formed, but it differs from other sands in being clay. It is green because of its colour. Under the microscope you can see it is composed of blue mud coloured by the presence of iron. The iron is usually oxidised to yellow, and it turns brown on weathering. Before weathering it is grey. I wonder why they called it greensand.   *T.N.G.*

This rock originated in a sandy dessert environment.   *N.H.*

A quartz wedge consists of a strip of mica.   *F.E.T.*

11.
THE FIRST MUDDY DIVISION IS THE OXFORD CLOG

The basic law of stratigraphy is the Law of Supposition.
C.J.B./H.E.W./D.K.B.

During the Ordovician, the seas osculated back and forth. *R.L.B.*

Chalk ... shallow sea ... not more than 4 to 4½ thousand metres.
J.H.McD.W.

Creataceous. *J.H.McD.W.*

Permian resting uncomfortably on Carboniferous Limestone. *G.W.T.*

The Silurian facies changes are well developed in the Silurian.
J.H.McD.W.

The only way of correlation is to follow the grits into the shales and the trilobites into the graptolites (one gradually gives way to the other).
J.H.McD.W.

Lludlow. *J.H.McD.W.*

The sediments change as the deposits go up the side of the geosyncline.
J.H.McD.W.

The Crustaceous System. *J.H.McD.W.*

A constituent of a coal-measure cyclothem is a muscle-band. *C.P.H.*

The Torridonian series was laid down in a hurry. *J.H.McD.W.*

Very difficult to correlate since the facies are triachronous or more. *J.A.*

The first geological event during the Tertiary orogenesis was the formation of a geosyncline. This occurred during the Precambrian. *C.J.B.*

Carboniferous trees fell down because they were soft and pithy inside.
C.P.H.

The Carboniferous is the most economic system. *J.H.McD.W.*

During the Jurassic conditions on land were probably deep seas.*C.J.B.*

There followed a period when the climate rose and the area became swampy. *C.J.B.*

Precambrian rocks are perhaps the most interesting [of Leicestershire] although not so economic as other parts of Leicester. *J.H.McD.W.*

During the Tertiary, psalms ranged farther north than at present.*R.L.B.*

The chalk of Cretaceous age is unbedded at the bottom but becomes more massive towards the top.   *C.J.B.*

This limestone was formed in deep seas in the Upper Old Age.   *C.J.B.*

The base of the Cambrian is dated at 6000 million years ago.   *C.J.B.*

[*Non-howlers, s.s.*] The Wearde grit — Devonian S.W. England, The Manor Hotel Beds — Devonian.   *C.J.B.*

There are five different kinds of limestone — oolitic, crinoidal, chalk, calcium carbonate and Jurassic.   *A.R.McG.*

Deep-Water faeces of the lower Paleozoic.   *J.A.W.*

The first muddy division is the Oxford Clog which is remarkably uniform followed by the Wenlock Limestone.   *J.A.*

The Upper Jurassic beds show the finest examples in the stratigraphical column, and the majority are used in industry.   *C.J.B.*

During the Trias the Pennines were eroded away, thus joining Lancs and Yorks and causing much bitterness in the lakes.   *A.R.McG.*

The seas of the Jurassic saw a change unparalleled in earlier times. The great saline seas that had formed the region known as Tethys began to evaporate, losing its sauces of replenishment.   *J.A.*

The comparison is that the rocks comprise two successive steps in a process, with a gradual transition from one to the other, and the contrast is in how different the two ages of rock-formation are, when the gradual transition cold-warm is omitted and the start and finish (so to speak) are put side by side.   *J.G.McD.*

The study of sediments unlocks the past, i.e. it is the key to Uniformitarianism.   *J.H.McD.W.*

The Limestone laid unconformidably on the older strata.   *C.J.B.*

The unconformity was laid down later than the original rock and so has suffered less erosion.   *C.J.B.*

The unconformity was caused by the Carboniferous limestone lying on folded Silurian slates. In this way the whole of the Ordovician was missed out.   *C.J.B.*

Duracic limestones.   *C.R.M.*

Renig volcanics.   *C.R.M.*

The Silurian is zoned by fossils.   *J.H.McD.W.*

The age of the Dalradian has been established by the use of fossil
mermaids, known as Aquatarts.   *H.P.*

Underneath the Proterozoic are unformable layers, which, how far they go down we do not know.   *R.L.B.*

Oolitic limestone, divided into the Inferior and Greater Oolite in the Cotswolds can be seen to be formed today on the Bahamas Bank.   *J.A.*

The Lower Greensand of the Weald is red in colour.   *C.J.B.*

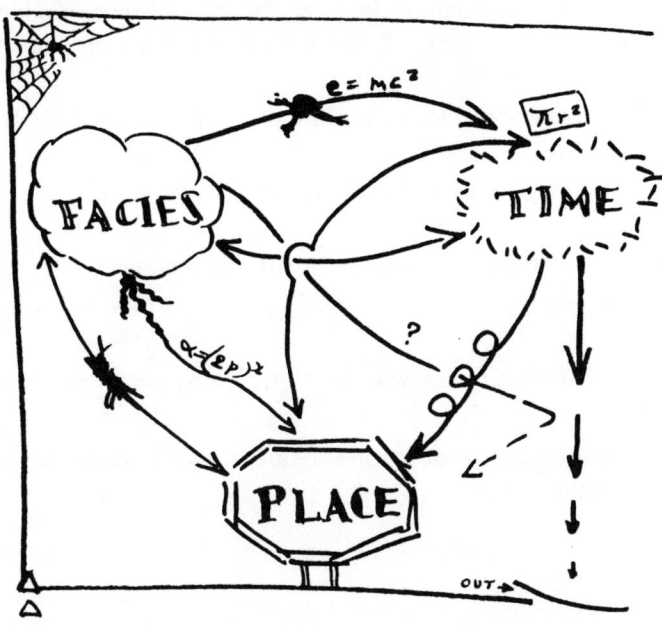

Facies is a term used when the same rocks are deposited in different places at different times, or when different rocks are deposited in the same place at different times or when different rocks are deposited at different places at the same time ....   *T.D.F.*

Cambrian time in Ohio was mostly under water.   *R.L.B.*

The sea went shallower and conditions went either desert conditions of to or to near desert conditions the sandstone (later quartz) was laid down on this.   *J.H.McD.W.*

The shelly facies is ejected from the vent and mixed with the graptolitic facies.   *A.R.McG.*

Oslo Caradocian — locally called the Pandy Ash.   *J.H.McD.W.*

The thin layer of shales and lsts. with fossilised oysters were layed down in shallow water, possibly during the Palaeozoic era, when molluscs were articulated. *G.F.*

Ammonites had a short history, mostly in the oredivision time period.
G.F.

In the late Palaeozoic, the Appalachians suffered a deeply moving erosional experience. *R.L.B.*

Two fingers of land came up from the SW, one of them proceeding up the Severn Estuary. The grits and sandstones built up and initiated the Great Armorican Orogeny. *J.A.*

The Law of Superposition: young things in beds are better than old things in beds. *W.H.W.*

The Phuket Group, peninsular Thailand: a Palaeozoic? geosynclinal deposit. *A.J.W.*

The climactic conditions were warm with an advance. *A.R.McG.*

This is a recent Tertiary and Palaeozoic fossil from the Gault Clay.
C.P.H.

The process of correlation was first started by Adam Smith who was an engineer. *J.A.*

The first characteristic of a zone fossil is that it is cecile; that is, it can move about freely. *C.P.H.*

12.
DIP HORIZONTAL, STRIKE UNKNOWN

The fault usually causes the oil to get lost, and the same with the dyke.
I.M.S.

[*Of Miocene saliferous beds:*] We wonder if the folding is worse than it looks or looks worse than it is? *S.E.*

[*'Walking' a prominent sandstone outcrop around a small conical hill, a geologist surprisingly found himself stratigraphically lower than his starting point:*] This must be spiral bedding. *S.E.*

Dip horizontal, strike unknown. *L.G.M.*

The fault caused the rocks to become nonconformist. *C.J.B.*

The faulting is flescular slip type. *A.R.McG.*

Anticlines & Sinclines are produced by squeezing upon a bed. *C.R.M.*

The Highland Bounder Fault. *G.B.*

[*Paperback Title:*] Crumpled Bedding — True Confessions of an Amorous Geologist. *H.P.*

Strata with a downward tendency form a sincline. *W.E.S.*

Beds lying uncomfortably on other beds. *W.E.S./G.W.T.*

These rocks form a sharp antisyncline. *A.R.McG.*

A horizontal fault is called a tear or wench fault. *C.J.B.*

The folding was on a north—south, east—west axis. *J.A.*

Beds of interminable thickness. *J.H.McD.W.*

The strike tells us the horizontal slope of the rock. *I.M.S.*

A fenster is a saddle on which a horst rides. *T.N.G.*

The distance AB is the primary creeping area. *P.G.C.*

There is flowage resulting in "neck and pitch" structure. *P.G.C.*

The crest point is defined as the highest point on the crest. *P.G.C.*

So originally the area was composed of an anticline and an antagonistic syncline. *N.B.*

Schistosity is the idea of mountains floating as islands on top of other rocks. This is the theory of a philosopher called Schist and is nowadays regarded as correct. *I.M.S.*

Now Mr. Holland followed the ordinary procedure of having tennis courts on the lawn at the back of his house, from which can be obtained a grand panoramic view towards the Chiltern Hills, which he built for himself 24 years ago. *D.P.*

The fault clearly sorts out the rocks in the formation. *D.F.B.P.*

A strike fault being a sideways movement will show all the right outcrops on the map but in the wrong places. *N.B.*

Being horizontal a sill stands a good chance of not appearing at all on a map. *N.B.*

A dip-slip fault is one in which the dip of the fault has slipped. *N.B.*

The general succession of the area can be seen by running from the north west corner of the map to the south east corner. *N.B.*

The cleavage in slate does not necessarily all have to be in the same direction. As it usually is not. This is due to the fact that when a bed of slate is folded the cleavage is formed then and can be in size or direction. *I.M.S.*

[*Research project:*] the study of stain in deformed socks.
W.W.B./A.E.S.M.

Bondinage structure. *P.T.C.*

Sinister tear faults. *A.M.E.*

The Highland Boundary Fault goes across Loch Lomond in the Bahamas area. *J.H.*

A fenster is a large lens to examine rocks through. *T.N.G.*

All that was left (after erosion) was the outriders on either side of the fault. *D.F.B.P.*

Fracture cleavage is not parallel to the axial plane but depends on the mythology of the rock. *A.D.W.*

The fault could be described as far from normal. *C.P.H.*

The fault is a recombinant type, having been once been a normal fault but then subdued by metamorphism. *C.P.H.*

The strike of an outcrop is the line at which when viewed vertically on a map or in the field is recorded by a compass as from due north. *G.B.*

In a normal fault the direction of the throw is from left to right, in a reverse fault the direction of throw is from right to left. *G.B.*

A normal anticline is one in which the folded rock has remained on the horizontal.   G.B.

The Highland Boarder Fault ....   G.B.

Rocks once like this   can come to look like this

                                                      J.A.

I tested both sides of the fault with dilute HCl acid to see which had been thrown down.   C.P.H.

At an oil-company exploration conference in the Middle East many years ago, some anticlines were described as being over 100 miles long — "twice the distance from London to Brighton." Voice from the audience: "But not half so nice."   S.E.

[*Spoonerism:*] Areas in which the fold axes run in the same general direction are known as homosexual actors.   H.P.

The detailed structure of this area avails me.   R.H.G.

The cause of intrication must be due to crustal shortening.   W.D.I.R.

13.
PHOSSILS

Fossils may be buried in hot material where they are burned to death
and preserved as amber.   C.D.G.

As it moved backwards the anus swallowed a genital plate.   T.D.F.

That well known Cretaceous fossil *Actinoclimax penis* ....   T.D.F.

Ammonoids are a present day fossil because there are no more living
ones.   I.M.S.

One fossilised animal is the rare trilobite, for although it has been
established when it existed, too many were never preserved for it to be
of any use in correlation.   C.D.G.

Abductor muscles of the brachiopods.   J.A.W.

The sicula would be razor-sharp to cut food.   A.R.McG.

Muscle bands occur in the Coal Measures.   J.A.W.

The incisors were fuzzed together.   GeoT.

Darwin's theory is based on three points — the struggle for exits,
survival of the fattest and maternal selection.   D.A.B.

The gross form of all these conifers is unknown, but present-day species
suggest the predominance of the arboreal habit.   W.D.I.R.

A mammoth in amber.   C.D.G.

Irregular echinoids can move in a straight line but regular echinoids
move in any direction in ever decreasing circles.   W.W.

An example of geographical speciation is the lesser Black Billed gull which can not breed with its brother the herring .... G.M.W.

Nautilord. J.H.McD.W.

By application of the fossil record and Darwin's collective thoughts, we can deduce what existed in the past. R.L.B./P.F.B.

Ammonites were floating creatures which when they were young had no coils but as they progressed in life they started to coil. When they died they uncoiled. D.F.B.P.

Trilobites occurred in the Carboniferous Period; we know this because they lived in trees. P.T.C.

Genus: *Teredo;* Range: Jur.-Cret.; Habitat: Bores into ships. N.H.T.

The hoof in the horse developed as the hand became the leg. GeoT.

Some large dinosaurs had three horns and were called triceps, others had two horns and were called biceps. T.N.G.

A brachiopod possesses a pedicle with which it can propel itself in a series of jumps. C.P.H.

The presence of genal spines, e.g. *Trinuclius,* is regarded by some as a sign of decadence. C.P.H.

The graptoloidea divulged themselves from the Dendroidea. C.P.H.

It is accepted that no animal can ever again exist after it has been declared as dead. C.P.H.

The trilobites that did not have eyes were, very appropriately, described as blind. C.P.H.

The animal oo built itself an exoskeleton of calcium carbonate which
is known as an ooid. *C.P.H.*

Clines — where 2 communities are separated by a geographical barrier
or just shear distance. *W.D.I.R.*

Articulate brachiopods have teeth and socks. *C.J.B.*

One of the fundamental differences in shell structure between
Inarticulates and Articulates was punctuality. *C.J.B.*

A dinosaur was large it had to eat night and day to keep itself from
starving. This did not give it time to eat its meals properly and it never
got enough sleep, so it became frustrated and defeatist and finally
extinct. *T.N.G.*

Trace fossils are used for tracing a sequence of beds from country to
country. *I.S.J.*

Trilobites and graptolites had very short lives (perhaps an era or two.)
I.S.J.

It appears that *Trinucleus* got fed up with not being able to see, so it
grew many eyes and a large genial spine to protect itself. *J.A.W.*

Pantheontology. *G.B.*

A dinosaur had a long tail which if you kicked it at the tip would leave
you time to run round to the front of the animal and laugh in its face
before it transmitted the shock to the brain. This led to its extinction.
T.N.G.

Reefs are what you put on coffins. *H.C.H.*

[*Of Didymograptus*] This fossil belongs to the family of corals, *i.e.* it
is a member of the crinoids. *A.R.McG.*

Flies have been known to have been preserved in minerals such as
Agerine, e.g. in Sweden. *J.A.*

An example of complete preservation can be seen in Russia where a giant sloth was found and remained intact down to the last detail. Even the food it was eating at the time was still preserved in it's mouth. This sort of example is extremely rare and there are perhaps only about 3 or 4 examples in the world. There have been several suggestions put forward regards how this happened. One such view was that a freik wind was the cause. This probably had a temperature of about -300°C which would freeze and kill anything instantaneously and it must have gusted anything up to 500 m.p.h. *J.A.*

Trachiobite. Crinoliths. *G.B.*

Radiolarian ooze: when radiolaria die their testes sink to the bottom of the ocean and build vast deposits known as radiolarian ooze. *G.T.*

Loess — a glacial flower of a plant which at one time spread over a great land area. *GeoT.*

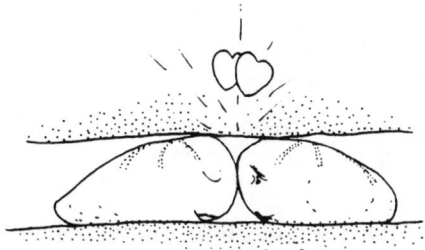

Echinoids love in burrows. *C.J.B.*

The Trilobite lived in Cambrian times at the bottom of the sea and was not very intelligent it had a very bad sense of smell and eyesight at a later date some trilobites developed without eyes because they lived in an area where it was so dark that they did not need them. *J.A.*

The facial suture became more advanced and lay behind the genital angle. *J.A.*

There are two basic concepts dealing with evolution with specific reverence to fossils. *P.F.B.*

... ponderosity of evolution. *P.F.B.*

Polyps swim about the sea when they are young, and when they get old they fasten themselves on their relations and live like that for the rest of their lives. *H.C.H.*

One of these animals of higher development is the chimpansy. *R.L.B.*

In the early Cretaceous, life for these reptiles became an acute struggle for extinction. *R.L.B.*

Devonian lung fish — If the wet season did not come they would have aided their fossilization by burying themselfs. That is probably why they are found in little groups, e.g. at Ledbury Station. *J.H.McD.W.*

... a carbonivorous fish. *P.F.B.*

A dinosaur is an extinct animal still found in Australia. It was sometimes so large that its feet are found in the Precambrian and its head in the Silurian because it was too big to lie down where it died. *T.N.G.*

The echinoids moved the anus and the mouth in various species as nature's way of catering for all possibilities. *J.D.L.*

Some trilobites had large eyes to enable them to escape from their creditors more easily. *N.H.T.*

The dodo is a bird that is nearly decent now. *H.C.H.*

Dr. Charles Darwin lived just long enough to receive the admiring tributes of the whale community. *D.P.*

They [echinoderms] have little competition in this diet (of corals). They eject their stomach out over the food and then suck it all back inside. *ConG.*

Echinoids move by jet propulsion; they rapidly excrete waste food through the anus and shoot forwards. *ConG.*

[Of graptolite reproduction:] I don't know how the little budders do it. *P.A.*

Dinosaurs had a large pituitary gland that led them into trouble and some theologians say to their downfall. *T.N.G.*

A skeleton is a man with his inside out and his outside off. *H.C.H.*

Genital spines (of trilobites) *N.H.T./J.A.W.*

The eurypterids that lived in the Old Red Sandstone were sometimes fifty times bigger than their nearest living relatives the scorpions, and their sting was fifty times as strong. Although they lived millions of years ago their sting has not completely ovaporated away, and it is still dangerous to collect fossil specimens of them near Lesmahagow. *T.N.G.*

The lamellibranchs reached a maximum in the Tertiary, but are now being overshadowed by the gastropods who have not yet reached their acne. *C.J.B.*

A dinosaur is a plant-eating mammal that preyed upon its fellow reptiles. It appeared in the Triassic period and died at a very old age in the Cretaceous. T.N.G.

The free cheeks of trilobites moved around and may have been used as some kind of steering device. C.P.H.

Ammonites went berserk in the Upper Greensand. J.A.W.

A dinosaur was fairly large, about 200 feet high. It balanced itself on its tail with the help of its hind legs. T.N.G.

The Ammonites were trapped in the limestone while looking for food. C.J.B.

.... chorals are formed on it which eventually turns to limestone. G.B.

The Carboniferous rocks are correlated by the fossil corrals present in them. G.B.

Some echinoids have an eccentric mouth or anus or both. C.P.H.

Trinucleus has a girder which is used for some ecological purpose like moving mud. T.D.F.

A dinosaur was a herbaceous animal that got so large it was too fat to run after its prey and so died of starvation. T.N.G.

The most salient feature of this new fossil is that it is petrified. M.S.

Phossils. J.H.McD.W.

A bone-bed is an infilled burrow of arthropods with their own bodies when they die. C.P.H.

Corals are sociable creatures who live in colonies.   *J.H.McD.W.*

Episcopalian trilobites.   *J.H.McD.W.*

Foraminifera are small, but they kept together and so are easy to see in rocks with the aid of a microscope.   *C.P.H.*

Trilobites were mainly marine animals but some were able to fly and others to walk on dry land.   *C.P.H.*

Finding of fossils enables the scientist to determine the age of the fossil and also the age of the hostess rock.   *GeoT.*

Conodonts have special problems.   *W.D.I.R.*

Graptolites are of interest to phalontolygists.   *I.S.J.*

There is the odd occasion when plants have almost been totally preserved without being fossilised.   *C.P.H.*

Coral reefs formed from *Globigerina*.   *J.H.McD.W.*

[Of Terebratula:] The fold in the commisure is a sign of decadence which would substantiate the evidence that the creature was sessile.   *D.E.G.B.*

Graptolites are often found in deep seat deposits.   *C.P.H.*

Shales bearing famous *Glossopteris* and *Gangamopteris* plant genie.   *J.H.McD.W.*

Fragment of holotype [sporangium] which could repay mass oration.   *W.D.I.R.*

In the articulate brachiopods the evolution of the species takes a somewhat different from what might be expected. Instead of dying out entirely some species gradually evolved into other species before actualiy dying out.   *C.P.H.*

Fossils are of general interest to the amateur geologist because they provide an interesting relief from other forms of geology and are nice to collect.   *J.A.*

In the Trinucleina we have a bazaar modification.   *A.R.McG.*

The creature has a stomach that can go outside the shell and eat its prey.   *W.G.J.*

A goniatite is a coughalopod — chest-footed.   *T.N.G.*

The leg consisted of 3 parts .... a projecting tooth-like device which projected into the stomach and masticated the creature's food as it walked. *A.R.McG.*

The gas chambers were used as a bouncy mechanism. *A.R.McG.*

The last whore of an ammonite. *A.R.McG.*

Mya would find it necessary to be moved since it has no means of motivation. *A.R.McG.*

The bivalves have a large opening at the anterior end which is used to extinguish water. *A.R.McG.*

On the dorsal surface of an echinoid is a plate called something, as is everything in this world. *A.R.McG./J.A.W.*

The sand is full of oysters, muscles and limits. *C.J.B.*

The early growth stage of the trilobite is the pteraspid. *A.R.McG.*

Graptolite life mode is hanging from possible trees into a river, feeding on flies and insects. *A.R.McG.*

Graptolites are called graptolites because they look like hieroglyphic writing and they are the origin of the Egyptian alphabet. *T.N.G.*

Of course, fossils can have lots of different shapes. *P.C.S.-B.*

Reef building corals are limited to coral reefs. *A.R.McG.*

[*Of ambulacra:*] The function of the pores is to close the shell in a zip-like fashion. *A.R.McG.*

Bioturbation is a method of staining biological structures in rock.
C.P.H.

Cast of a fish with little imagination. *J.H.McD.W.*

Trilobites often had a mucronate tail which they used for anchoring themselves into the mud when feeding. They probably posed rather like a cobra when feeding on microphagons, diatoms etc. *J.A.*

The gastropods are of the greatest importance showing this, evolving from early Palaeozoic marine molluscs to estuarine, terrestrial and even flying forms. *J.A.*

Dinosaurs were so heavy that their legs were always near the ground. This was a fault that led to their extinction. *T.N.G.*

[*U.S. newspaper:*] Dr. Jones's argument for believing that there are countless other worlds where living beings are present, briefly, is this:
  Ninety per cent of the shrimps served on the tables of the United States come from the coastal waters of Alabama, Florida, Georgia, Louisiana, Mississippi and Texas. *D.P.*

The bryozoan has a separate anal opening in addition to the mouth. The coral uses its mouth for both. *R.L.B.*

Trace fossils, such as dinosaur footprints, may also be used. A dinosaur would not walk, for example, on the roof of a cave, so a bed is overturned if footprints appear on the rocks of the roof. *M.F.*

If footprints are found on the underside of a bed it usually means the bed is inverted as few animals walk upside down underground.
J.H.McD.W.

Zaphrentis fossils are often found in deep water shales because when they died they toppled over and rolled off the continental shelf. *C.J.B.*

Photoplankton. *J.H.McD.W.*

A brachiopod has an operculum and an eye plus a foot to keep it stable in the ground. *N.J.H.*

It is much more abundant in species with ammonites of the Jurassic becoming highly specialised ready for extinction. *J.A.*

If trilobites are found on the bottom of a lake then it is concluded that the lake was not there years ago, as trilobites are not marine living.
B.J.B.

When the water rose the island which was already submerged could not rise but the corals could and did. *J.G.McD.*

Originally, there were huge forests; prehistoric animals destroyed them and slowly trampled them into the soft mud, forming coal. *C.D.G.*

Petrifaction is what the name suggests. Animals when given shocks or frightening experiences seem to "freeze" on the spot. This is called petrifaction as they are said to be petrified. *I.M.S.*

No animals existed but large extravagant Insects were found [Coal Measures]. *J.G.McD.*

Fossils in S. Wales Millstone Grit are Llamelibranchs.
A.R.McG./J.H.McD.W.

Both colonial and simple Productids were seen. *C.R.R.*

The brachiopods I personally observed had a length of about ½" this being evidence to the marine origin of this rock. *C.R.R.*

Fairly well defined choral bands are also visable. *C.R.R.*

Oysters are bi-valved shellfish belonging to the graptolites. *G.F.*

Oysters are brachiopods. *G.F.*

Toledo is an example of a wood-boring bivalve. *A.R.McG.*

A gastrolith is the gall stone of a reptile. *C.J.B.*

[Behavioural howler] Frustrated dinosaur-watchers wait for hours in Dinosaur National Monument, Utah, waiting to glimpse a dinosaur crossing the trail. *W.D.I.R.*

Crinoid forests of the Carboniferous. *J.H.McD.W.*

Professor versteinert von fossilen Illusion. *F.F.*

Thecal elaboration culminates in the sophisticated **insularity** of *Rastrites*. *W.D.I.R.*

Millipede trails occur, along with body fossils, apparently in fishes in the local larvas. *W.D.I.R.*

Terebratula has a circular hole in the beak, out of which waste material is ejected. *D.E.G.B.*

A facial suture cutting the posterior border of the cephalon is termed improperia. *J.A.W.*

CITY OF GLASGOW DISTRICT COUNCIL.
DEPARTMENT OF CLEANSING.

DINOSAUR CONTAINERS.

OFFERS are INVITED for the SUPPLY of 7 only 25 cu. yd. CONTAINERS and 3 only 35 cu. yd. CONTAINERS.

Specifications and forms of tender, obtainable from the Director of Cleansing, 235 George Street, Glasgow, G1 1QZ, to be returned to me by 31st October, 1979.

STEVEN F. HAMILTON,
Director of Administration and Legal Services.

City Chambers,
Glasgow, G2 1DU.
15th October, 1979.

F.W.

CONTRIBUTORS

J.A.	J.Addison	T.N.G.	T.N.George
P.A.	P.Allen	ConG.	Con Gillen
B.W.A.	B.W.Anderson	R.H.G.	R.H.Graham
E.C.A.	E.C.Appleyard	C.D.G.	C.D.Gribble
D.K.B.	D.K.Bailey	J.H.	J.Hall
P.F.B.	P.F.Ballance	N.J.H.	N.J.Hancock
M.B.	Michael Bamlett	F.H.	F.Hodson
M.J.L.B.	M.J. Le Bas	N.H.	N.Holgate
D.A.B.	D.A.Bassett	A.M.H.	A.M.Hopgood
D.B.	Denis Bates	R.A.H.	R.A.Howie
R.L.B.	Robert L. Bates	C.P.H.	C.P.Hughes
W.W.B.	W.W.Bishop	H.C.H.	H. Cecil Hunt
B.J.B.	B.J.Bluck	W.G.J.	W.G.Jardine
G.B.	George Bowes	I.S.J.	I.S.Johnston
D.E.G.B.	D.E.G.Briggs	J.D.L.	J.D.Lawson
M.Br.	M.Brooks	G.L.	G.Lea
Ga.B.	Gavin Brown	B.E.L.	B.E.Leake
C.J.B.	C.J.Burton	J.G.McD.	J.G.MacDonald
N.B.	N.Butcher	A.R.McG.	A.R.MacGregor
J.C.	J.Campbell	R.M.McI.	R.M.McIntyre
P.T.C.	Peter T.Carr	A.C.McL.	A.C.McLean
P.G.C.	P.G.Cooray	A.E.S.M.	A.E.S.Mayer
K.C.	Keith Cox	C.R.M.	C.R.Morey
D.L.D.	D.L.Dineley	L.G.M.	L.G.Murray
(and members of the Bristol University Geology Dept.)		D.F.B.P.	D.F.B.Palframan
		H.P.	Henry Pantin
G.P.D.	G.P.Durant	D.P.	Denys Parsons
S.E.	S.Elder	W.S.P.	W.S.Pitcher
A.M.E.	A.M.Evans	I.P.	I.Price
R.C.E.	R.C.Evans	W.D.I.R.	W.D.Ian Rolfe
G.F.	George Farrow	C.R.R.	C.R.Rowley
M.F.	Michael Fleuty	J.S.	J.Saxon
T.D.F.	T.D.Ford	T.C.F.S.	T.C.F.Sibly
F.F.	F.Frölicher	I.M.S.	I.M.Simpson
GeoT.	*Geotimes*	D.G.S.	D.G.Smith

F.S.	Sir Frederick Stewart	G.P.L.W.	G.P.L.Walker
M.S.	Marie Stopes	J.A.W.	J.A.Weir
W.E.S.	W.E.Swinton	W.W.	W.Welsh
P.C.S-B.	P.C.Sylvester-Bradley	W.H.W.	W.H.Wheeler
F.E.T.	F.E.Tocher	J.H.McD.W.	J.H.McD.Whitaker
W.E.T.	W.E.Tremlett	A.J.W.	A.J.Whiteman
N.H.T.	N.H.Trewin	F.W.	F.Willett
G.T.	G.Trollope	H.E.W.	H.E.Wilson
G.W.T.	G.W.Tyrrell	G.A.W.	G.A.Worrall
G.M.W.	G.M.Walkden	A.D.W.	A.D.Wright

Lothar Deubot.
Feb 1982.